DEVELOPING SKILLS IN COMPUTER SECURITY THROUGH LEARNING'S EVALUATIVE OBJECTS

DOUGGLAS HURTADO CARMONA

DEVELOPING SKILLS IN COMPUTER SECURITY THROUGH

LEARNING'S EVALUATIVE OBJECTS

Dougglas Hurtado Carmona

© **2012, Copyright in this edition:**
Dougglas Hurtado Carmona
dougglash@yahoo.com.mx
dougglas@gmail.com

ISBN: **978-1-4716-5786-3**

Translated from:

Desarrollando habilidades en seguridad informática

por medio de objetos evaluativos del aprendizaje

ISBN: 978-1-4716-5760-3

Copyright © 2012, Dougglas Hurtado Carmona

ACKNOWLEDGMENTS

GOD Almighty.

The students, which very vividly, unknowingly participated in this project

AUTHOR

DOUGGLAS HURTADO CARMONA

The author is a Systems Engineer, Master in Computer Engineering graduated from the University of North in Barranquilla, Colombia. It has also supplemented his studies with complementary studies in Minor in Management and Information Systems Security, Bureau Veritas Certification to ISO27001 ISMS Internal Auditor. Diploma in Scientific Research, Development for Web Applications, IT Security and Computer Forensics Education and Pedagogy.

In your professional role has been highlighted as International speaker, computer security engineer, Project Manager of software development, systems analyst and programmer, IT project manager, independently advised companies involved in building software.

With respect to the educational part whatever with 14 years of university teaching experience in the areas of computer engineering. And is a researcher on the topics of Computer Security, Computer Forensics, General Systems Theory and system dynamics to software engineering and theory of compilers.

Within the field of research and development, research highlights:
- "Analysis of the development of competencies from the use of Computer Assisted Learning," which received Special Mention in the ACOFI AWARDS 2007
- "Methodology for the development of systems based on learning objects"
-

Other books include:
- General Systems Theory: A focus on computer science engineering
- Analysis of skills development from computer-assisted teaching

He currently directs the Deanship of the Faculty of Engineering of the University Foundation of San Martin, headquarters Puerto Colombia, in the city of Barranquilla, Republic of Colombia.

TABLE OF CONTENTS

Chapter 1

ABOUT THE INVESTIGATIVE PROJECT

INTRODUCTION

Within the range of options to strengthen the various learning processes in higher education worldwide, has occurred in recent years, the use of technologies of Learning Objects (LO) [Wiley, 2001], and especially in what refers to independent work to be performed by student in the process of professional training.

Very even though the learning object technologies are on track to reach a mature stage in its development [Friesen, 2001], its usefulness in the education process at all levels and disciplines is undeniable, and especially valuable.

Among most significant contributions are found its contribution to increased development of competencies (knowledge, know-how and do) to a minimum of 30% when the student benefits from a learning process that is structured to the assistance of ICT technologies. [Hurtado Carmona, 2011a]

Following a path a bit more technical on the basis of the modular structure of learning objects [Wiley, 2006] and the concept that learning objects can be considered as systems [Johansen, 1996], and especially the concept of "set of parts (subsystems) that interact to accomplish a goal" [Hurtado Carmona, 2011b], can delegate the function of assessing the skills acquired in a learning process, exclusively to a part or subsystem, or a complete learning object, depending on the complexity of the evaluation. These objects or subsystems, called Learning´s Evaluative Objects (LEO) [Hurtado Carmona, 2010c].

The evolution and functional specialization of learning objects is accomplished with the birth of Learning's Evaluative Objects. It is in this context where this project is developed, aiming to describe the functionality of Learning's Evaluative Objects, and the results of its use in the teaching of the subject of computer security.

This document is an extract of the most important aspects of the research developed, which is presented as follows: First of all, sets out some basics concept of learning objects and Learning's Evaluative Objects. Then, presentation and brief description of the problem, then work goals, immediately, methodological aspects, later describes the results obtained from the use of Learning's

Evaluative Objects on the subject of computer security and computer forensics undergraduate and graduate, finally set out the conclusions and possible future work.

FUNDAMENTALS

Learning Objects

In general is considered a Learning Object "...elements of a new type of computer-based education grounded in object-oriented paradigm of computer science. Object orientation highly values the creation of components (called objects) that can be reused..." [Wiley, 2006]

With the above and the need for the learning object possesses the qualities to be searched, retrieved and reused in different scenarios, it becomes necessary that the learning object is described through a set of *Metadata*.

An essential attribute of learning objects is size, since this attribute is suited to be used as part of a lesson or module. Additionally, it must also be reused, which is nothing to possess the ability to be used in different units or learning activities. Important is that the learning object is accessible, and that benefits the learning process that can easily locate and usability. The characteristic of being durable is reflected in its maintenance should be low. Interoperable, is can use it in various technological platforms, or different course management systems [Arsham, 1995].

A summary definition of learning objects would be: "... A digital entity that allows a pedagogical process of a bare minimum of training content that engages the target, the development, implementation and evaluation..." [Hurtado Carmona, 2010c].

Learning's Evaluative Objects

It is defined as Learning's Evaluative Object (LEO) to *digital entity whose function is to evaluate the interpretative, argumentative and purposeful skills about a subject no matter what the student has made the learning process* [Hurtado Carmona, 2010c].

As a fundamental feature, Learning's Evaluative Object when formulating problems or questions should be unpredictable and protects the integrity of the evaluation. Similarly, a Learning's Evaluative Object can be an independent digital entity, or be a component of a learning object, which has been entrusted with the evaluation, exclusively.

Among his closest antecedents, while far away in terms of its conception (second cousin), we have evaluative objects, worked by Vitturini [Vitturini et al, 2005], that they are a special type of learning objects, which in essence, make an entry constituted, where appropriate, by practical exercises solved by students, and output sorted into two groups, one that exceeds the minimum requirements and that should be delivered again, according to the criterion of correctness set by the teacher.

The structure of an learning`s evaluative object, consists of two engines, one, generator problems or questions, and another, in charge of the competency evaluation. Additionally, it has a module in charge of monitoring events and storing monitoring data in the encrypted file that contains the responses evaluated. [Hurtado Carmona, 2010c] Are described below:

1) *Engine generator problems or questions.* Comprised of algorithms that uses a database to dynamically generate questions and issues consistent. These algorithms must ensure that when making multiple assessments generated sen different questions each time or at least the number of collisions is very low. This engine is not simply having a bank of questions and take some at random, but to generate questions in a consistent manner.

2) *Engine skills assessment.* This engine has an algorithm that tracks and calculates the results obtained by the student at the quiz, and routines that are responsible for evaluating the skills associated with the student as it was in the process to address these problems.

3) *Event monitoring module.* It consists of a set of routines responsible for encrypting the evaluation and monitoring of student activities, to meet the test.

PROJECT DESCRIPTION

Project title

This research work has been titled with the name of: Developing skills in Computer Security, through *Learning's Evaluative Objects.*

Summary

This study describes a special type of learning objects, called Learning's Evaluative Objects (LEO), whose fundamental role is to assess the competencies and skills acquired through different learning options.

Simplified, first, detailing the structure and functionality of Learning's Evaluative Objects, and secondly, analyze the results obtained when using a Learning's Evaluative Objects in the area of Computer Security.

Keywords

Development of skills; Computer Security; Learning's Evaluative Object; Learning Objects.

Interested entity

Directly concerned entities correspond to the Faculty of Engineering, University Foundation of San Martin headquarters Puerto Colombia, Barranquilla, Republic of Colombia, and the Specialization Program in Computer Security, the University Research and Development-UDI the city of Bucaramanga, Republic of Colombia

Furthermore, this project may be of interest to individuals and institutions that are involved with the processes of evaluation and / or the development of skills in the area of computer security and other areas.

Estimated time

The estimated time for experimental realization of the project corresponds to nine (9) academic semesters from the second semester of 2006 until the second semester of 2010, within which information is processed and proceeds to document the results.

RESEARCH PROBLEM

Brief Description of the Problem

Since its inception in mid 2005, the course of deepening called *Minor in Computer Security and Computer Forensics*, which imparts to students of the Computer Engineering program of the University Foundation of San Martin headquarters Puerto Colombia, presented a "evaluation gap" in trying to assess the special skills and abilities acquired by students in some of the subjects who composed it.

In simple words, the conventional skills assessment was clear and well directed, but the problem was how to motivate the student, without giving the instruction when it is evaluated at different creative solutions breaking traditional patterns, this being necessary to the great differentiator of a successful professional in the area of information security.

This may be the product, among other situations, the following:

Castration of creativity in students at an early age: Oddly enough, the creativity of a person is mutilated by the type of education they receive from an early age, where they are taught to students to "think" in accordance with the guidelines of the school, family, country ; repressing so innate creative impulses. By limiting creativity, ensures that the institutions and models do not collapse. Recall the case of Galileo Galilei.

Falta de autoestima de estudiante frente a sus obligaciones: Un bajo autoestima y un sentimiento de "no creer en mi mismo" es el origen de hacer las cosas por salir del paso y no con verdadera vocación. Poco a poco la falta de auto estima genera una flojera y falta de bríos en el estudiante a la hora de realizar las actividades propias en su proceso de aprendizaje.

Lack of self-esteem of students meets their obligations: A low self-esteem and a feeling of "not believing in myself" is the origin, to make things out of trouble, not true vocation. Slowly, the lack of self-esteem creates a softness and lack of vigor in the student, when carrying out the activities in the learning process.

Indeed, the engineering student (and other professions) to be private, to enhance their creativity in solving problems, would be disadvantaged in reference to job performance.

Problem Formulation

This project seeks to answer the following question:

How to motivate the student's creativity to be assessed through Learning's Evaluative Objects to develop skills that allow him to propose solutions that break with traditional patterns in the Computer Security area?

JUSTIFICATION

This project aims to analyze the use of Learning's Evaluative Objects in the evaluation of skills development the area of Computer security in higher education students, as well as related aspects.

Also, this project plans to student motivation through Learning's Evaluative Objects, seeking to develop their own skills, which enable them unusual creative solutions in the Computer Security area.

This analysis of the results, will allow academics, and especially teachers, to show the advantages of the use of Learning's Evaluative Objects, to integrate this component to your educational culture in order to motivate the students' creativity, thinking about their professional development.

Another aim is to encourage the creation and use of Learning's Evaluative Objects in different disciplines.

OBJECTIVES

The general objective is intended to meet in the present study is stated as follows:

Analyze the use of Learning's Evaluative Objects in the assessment of skills development the area of computer security in higher education students in order to encourage creativity in them to develop skills that will enable it to propose solutions that break with the patterns traditional.

In order to tackle with the overall goal described above must meet the following goals:

- *Define the Computer Security topics that will serve as the basis for the realization of Learning's Evaluative Object.*
- *Building the Learning's Evaluative Object to be used.*
- *Design data collection instruments.*
- *Select the experimental sample.*
- *Apply the information collection instruments to the selected sample.*
- *Analyze the results to make graphs illustrating*

PROJECT HYPOTHESIS

Type of Hypothesis

Given that this project is framed in observing, in part, the behavior of students to be motivated to use their creativity when they are evaluated by the Learning's Evaluative Object, and complementary to perceive and to determine the influence of the use of creativity in solving problems in the development of skills in the area of Computer Security, its can certainly say that the type of hypothesis is causal.

Statement of Hypothesis

In the framework of objective sought with this investigation, it is necessary to know if you can accept the following hypothesis:

H1: *The motivation for the use of creativity through the use of Learning's Evaluative Objects influences the development of skills and abilities that enable the student to propose nontraditional solutions to problems in the Computer Security area.*

VARIABLES

Description of Variables

In this project, the following variables are proposed: *Using creativity* and *skills development*, which are described below:

Using creativity

This variable, *Using Creativity*, represents the student's determination to use the creativity to solve problems. This variable exhibits behavior such as "influences in" that defines its independent character. Its dimension is **assessment in the Computer Security area**. Presents a single indicator called the **Usage**, which takes Boolean values (True or False).

skill development

This variable describes the status, performance knowledge, skills and values resulting from the learning process of a professional activity in information security. The variable skill development has three (3) dimensions, namely:

- *Interpretative*: Refers to achieve success based on ability to make sense from either a text, of a proposition, problem, etc.

- *Argumentative*: Based on achievement gains, with guidance to account for a statement, articulate concepts and theories to support, justify, build relationships, demonstrate and conclude.

- *Propositive:* Grounded in achievements such as proposing hypotheses, solve problems and construct alternative solutions.

The dimensions of the variable named Skills Development has two indicators, the first known **Value** and the second called **Rate**.

- The indicator called **Value**: Presents positive decimal values.

- The indicator called **Ratio**: Presents decimal values within the range of 0 to 1, which are the result of the division between correct hits and quantity of evidence. The ranges of the quality assessment of this indicator are:

  Poor: [0%-59%]
  Acceptable: [60%-79%]

Good: [80%-90%]

Excellent: [91%-100%]

Operationalization of Variables

In the Table 1 is described the process of operationalizing the variables taking into account its dimensions and performance indicators:

TABLE 1. **OPERATIONALIZATION OF VARIABLES**

Variables	Dimension	Indicators
Using creativity	Assessment in the Computer Security area	Usage
Skills development	1. Interpretative	Value Rate
	2. Argumentative	Value Rate
	3. Propositive	Value Rate

DESIGN METHODOLOGY

Adopted design

The research design is **experimental**, since deliberately, it is induced to the establishment of the phenomenon being studied, this by creating the need for the students to find other ways to solve when it is evaluated (activation of the independent variable: **Using creativity**) and also is controlled environment and everything that can influence the phenomenon, and all, in order to observe the behavior of the dependent variable, called **Skill development**.

Research type

Since the claims of this project, which lie in obtaining basic knowledge and principles, oriented to establish a starting point for troubleshooting, we can say that the Research type is **Basic**

Information collection techniques

Techniques of collection of primary information

The source of collection of primary information to be used in this project is a **Learning's Evaluative Object**.

Description of the instrument used

The Learning's Evaluative Object used is called **OEASegInf** and was developed under the platform .Net, using visual user interface inspired in LCARS interface (fictional operating system of the Star Trek spacecraft and popularized by lcarscom.net franchise). A look to the interface of the Learning's Evaluative Object can be appreciated in Figure 1.

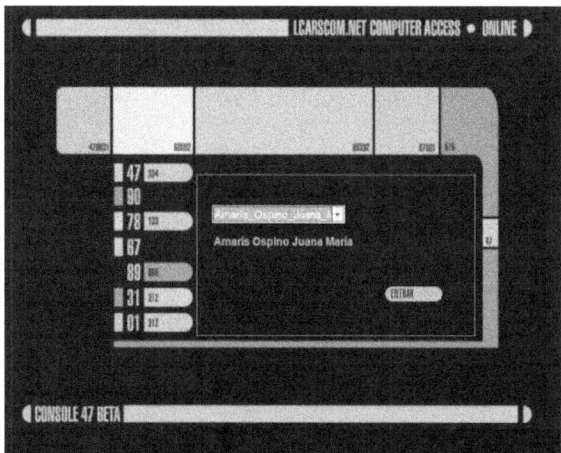

FIGURE 1. User selection screen of OEASegInf

OEASegInf is structured, first, by a *generating motor of problems*, which is based on the laws of Morgan, and in rules of semantic syntactic verification in respect to the statement of the problems. Being able to create 7x20x6 problems; is structured, first, by a generating motor of problems, which is based on the laws of Morgan, and in rules of semantic syntactic verification in respect to the statement of the problems. Being able to create 7x20x6 problems; constructing to questionnaires of 9 questions of $_{840+9-1}C_9$ combinations, where $_nC_k = n! /(k! (n-k)!)$ [Johnsonbaugh, 2005].

In second instance, conforms to the instrument, *a engine of evaluation of competitions*, which is in charge to register the answers of the student, evaluates them, storing them the information collected in an encrypted file; and in addition, with the purpose of to register the development of the additional propositive competitions, also it registers, among others, the following events:

- *Screen capture (screenshots with printscreen).* Some students capture the screen to keep an image from the questions that the LEO does to them. Lamentably for them the probability of questionnaire repetition is very low.

- ***Attempts of Cracking of software***. It is based on the fact to modify the binary code of the instrument so that it evaluates any answer that is selected like correct. Nevertheless, the instrument owns a method that is able to say if it were victim of this technique or no.
- ***Previous attempts of solution.*** Of equal way, the instrument registers the previous solutions and the times that has been realized the test.

Completes the structure of the Learning's Evaluative Object with *Event Monitoring Module*. As its name implies, this module has the task of monitoring the events and to generate a file containing the identification of student, responses to examination, and event monitoring data. This module uses, in turn, an encryption method to protect the information in the file, and additionally places a signature of protection to increase the security of the file, because if it can decrypt the file contents (break encryption) and create a dummy file with no real information, there is the option to verify the signature protection, simply put, there are two layers of security.

Below is an example of the contents of the encrypted file.

```
Æ´òN©®«µí-.<ºzbzEi• :dxü±´~¿w=†1ï^•• º¦• K0»ªKÇÃIÞ"]£Úœ€"2DU·C€+66's3:ÉááÒ„´_¢ŒÅÈRž×ÅÍŽq• ×Á
ÂiÀY6-di÷¹ÅN<• Ïþsù-f¼£¨»wP9Ÿ>T§ŸÄ¿ZŽ¼æ¼iÛÔG¢~f€éiKjâ¯¿p6†p›ƒÅ¥²ÔËdŽÔ-
+¤â'Û>¥ÃVõe2ZUá¥3¼ÞŸ¨ïÉ_š`¬s†XãuiW⬚MyäŠIÀk@¥"î¥•òP^v0OÀòÂH-
¦_qa¶é:aXË@È‰‰‰‰‰‰‰‰‰‰‰‰‰‰‰‰‰‰‰‰‰‰‰‰‰}ƒ{|...{~||„l}~†◻ †◻l¼z¹z‰‰
‰‰‰‰‰‰‰‰‰‰‰‰‰‰‰‰‰‰‰‰‰‰‰‰‰ÈYV'»¾¹~YVÈ‰‰‰‰‰‰‰‰‰‰‰‰
‰‰‰‰‰‰‰‰‰‰‰‰‰‰‰‰‰‰‰‰‰‰‰‰‰‰‰ÈYVl§œž•š Ÿ• ž"š©§˜±²À•  „µ¯·©§œž•š Ÿ• ž"š©§˜±²À•  „µ¯·©§œž•š Ÿ• ž"š
©§˜±²À•  „µ¯·©§œž•š Ÿ• ž"š©§˜±²À•  „µ¯·©§œž•š Ÿ• ž"š©§˜±²À•  „µ¯·©§œž•š Ÿ• ž"š©§˜±²À•  „µ¯·©§œž
•š Ÿ• ž"š©§˜±²À•  „µ¯·©§œž•š Ÿ• ž"š©§˜±²À•  „µ¯·©§œž•š Ÿ• ž"š©§˜±²À•  „µ¯·©§'• ©
```

Using the instrument (Learning's Evaluative Object)

The purpose of the use of the instrument is to evaluate the student who is pursuing a course in which the issues were given corresponding to the propositional skills to be evaluated, regarding the issue of computer security, whether in the course of deepening (undergraduate) or specialization (graduate). Together to evaluate the interpretative and argumentative skills, it is assigned reading or documentation on the issues associated.

The procedure for using the instrument is as follows: First of all, the student should study the documents assigned to prepare for the test. The instrument is downloaded from website designed for it (can be a learning platform such as Moodle) or sent by email. Next, the student installed on your computer, and performs the test as often as desired. Then when deemed to have made the best of all it sends the (those) file (s) of responses generated by the LEO to the teacher. Usually the student is requested to send five (5) output files at your discretion. We must clarify that the Learning's Evaluative Object does not tell anything about its assessment the student is limited to only report that has created a response file to be sent to the teacher.

So insistent is inserted in the environment of the evaluation, a high stress level, to try to generate propositive efforts of students, specifically for students to use the techniques taught in class. This stress is accomplished by assigning a very limited time to answer each question, which ranges in 30 seconds. When this is completed time is passed to the next question, and without realizing the test is completed in less than five (5) minutes. See Figure 2.

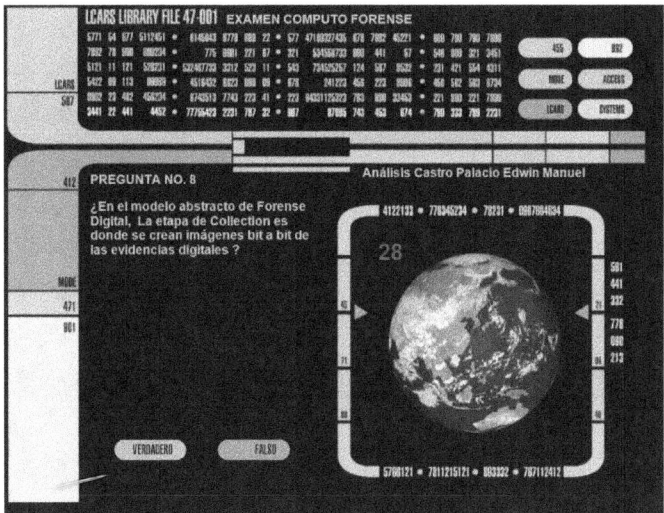

FIGURE 2. Assessment screen

Objective of the OEASegInf

Basically the objective of the Learning's Evaluative Object is not to answer questions, else, find a way to gain time to answer the questions properly, or rather, circumvent security of the Learning's Evaluative. This is the result of instrument design, because the questionnaire is designed in such problems so it is very difficult to answer it without the application of a technique studied in the course, which ultimately is to encourage student creativity to solve problems.

Competencies to evaluate

In accordance with the guidance of the evaluation process adopted, which focuses on the skills of the type interpretative, argumentative and purposive. The skills assessed are intended to interpret, analyze and articulate the concepts, tools, techniques and countermeasures are necessary to protect an organization's information to any digital attacker. Additionally, propose and try to use the attack techniques in order to evade the controls on the evaluation test.

The exercise performed is seen, first, deals with three essential techniques to evaluate the interpretative and argumentative skills are: Follow the trail, scanning and enumeration. These techniques are described below.

1) *Follow the trail*. It is based on gathering information about the goal of an attacker. Create a complete profile of the security policy adopted in an enterprise. Using a combination of tools and techniques, attackers can acquire an unlimited amount of information and diagramming it in domain names, network blocks, individual IP addresses and systems connected directly to the Internet. This procedure is necessary to ensure a systematic and methodical identification of any information relating to the technologies being. [McClure et al, 2010]

2) *Exploration*. Exploration is equivalent to identify the systems that are active and available to any attack. With the techniques used in the exploration, allows attackers to determine the systems that are active and which can be attacked more easily

3) *Enumeration.* The Enumeration involves active connections to systems objectives, with the respective structured requests to them. Hence, the information that an intruder looking for with the enumeration can be classified into the following categories: Resources Network and Sharing, Users and Groups, Applications and Messages. [McClure et al, 2010]

4) *Models procedure of Digital Forensic investigation*. Description of procedures for digital forensic investigation. Presentation of the different models of inquiry procedure. Description of your stages, advantages and disadvantages, and other items associated. [Baryamureeba and Tushabe, 2004] [Yong-Dal Shin. 2008]

Additionally, selected as additional competencies to assess some types of propositive, namely: Cracking software, social engineering and cryptanalysis.

5) Software Cracking. It consists in modifying the software to remove or outwit the methods of protection against unauthorized copying, changing an evaluation version (trail / demo) to complete, among others. Among the techniques used in the software cracking, include the disassembly and reverse engineering, which are of great help in modifying the binary code instructions for the software do what the attacker wants. In essence, this technique is to reprogram software.

6) *Social engineering*. Social engineering is the art of persuasion, with deceits, to people for that, almost without realizing it, give vital information to a stranger. [Kevin Mitnick, 2006] It is also considered, as all behaviors useful for getting information from people close to a computer, without them realizing they are revealing valuable information that compromises the security of a system. Social engineering is a very effective form of hacking, as it has a near 100% effectiveness.

7) Cryptography. Cryptography is known as the art of secret communication, where it transforms a plaintext (message), by means of a function parameterized by a key, resulting in a ciphertext (Cryptogram). This set is called a cryptosystem.

The Cryptanalysis in turn, is responsible for 'breaking' the cryptosystems. It uses different techniques to find the plaintext corresponding to a specific cipher [Pino Caballero, 2003]. Cryptanalysis often not dependent on knowledge of the encryption algorithm, but rather depends on the systems of mathematical approach that can discover the plaintext or the key. Therefore, the difficulty of cryptanalysis process depends solely on the information available, where, less information more difficult in the process of cryptanalysis.

Population and Sample

The population constitute the students who completed the course entitled systems and methodologies of access control, both the course of deepening Information Security and Computer Forensics of the University Foundation of San Martin headquarters Puerto Colombia (FUSM) in the city of Barranquilla, Colombia, and the Specialization in Computer Security of the University of Research and Development (UDI) in the city of Bucaramanga, Colombia.

The sample was made up to 404 students is constituted as follows:

TABLE 2. **ESTRUCTURA DE LA MUESTRAS**

PERIOD	INSTITUTION	AMOUNT
2006-2	FUSM	53
2007-1	FUSM	45
2007-2	FUSM	36
2008-1	FUSM	21
2008-2	FUSM	22
2009-1	FUSM	19
2009-2	FUSM	27
2010-1	FUSM	20
2010-2	FUSM	17
2008-1	UDI	20
2008-2	UDI	25
2009-1	UDI	24
2009-2	UDI	27
2010-1	UDI	23
2010-2	UDI	25

Information Processing

For the Information Processing shall take into account the following considerations:

1. All students enrolled in each semester will be taken as part of the sample.
2. The instrument will be applied to every student in the sample.
3. After obtaining the data are classified and tabulated
4. Procedure is used for hypothesis testing.
5. The results are shown in graphical form.

DELIMITATION

Conceptual delimitation

The theme addressed in the experiment refers to the topics of computer security in particular the following items described in Table 3:

TABLE 3. **CONCEPTUAL DELIMITATION**

COMPUTER SECURITY BASICS [Alexander, 2007] [Mann, 2011]
• Fundamentals of Computer Security • Policies for data security • Physical, logical and locative security
SYSTEMS AND METHODOLOGIES OF ACCESS CONTROL [Durán, 2010] [Hadnagy, 2011] [Zemanek, 2004] [Long, 2004]
• Basics of Access Control • Access control telecomputing • Control access to databases • Access Control by Social Engineering • Access control software • New trends in access control
COMPUTER FORENSIC [Cano, 2009] [NIJ, 2001] [NIJ, 2008] [Casey, 2009] [Baryamureeba and Tushabe, 2004] [Yong-Dal Shin. 2008]
• Introduction to computer forensics • Models of computer forensics • Techniques for gathering information and digital evidence • Tools in computer forensics
CRYPTOGRAPHY [Maiorano y Fernández, 2009] [Caballero Pino, 2003]
• Basics of cryptography • Symmetric Cryptography • Asymmetric Cryptography • Processes and procedures Cryptanalysis

NETWORK SECURITY [Zemanek, 2004] [McClure et al, 2010] [Hadnagy, 2011]
• Attack Techniques
• Defense techniques
• Techniques to trace, scanning and enumeration
• Security and software applications
• Social engineering

Delimitation Spatial

This project was carried out at the Faculty of Engineering of University Foundation San Martin Headquarters Puerto Colombia (FUSM) in the City of Barranquilla, and the University Research and Development (UDI) in the city of Bucaramanga, Republic of Colombia.

Chapter

2

RESULTS OF INVESTIGATIVE PROCESS

PRESENTATION OF INFORMATION COLLECTED

The tests were performed at a total of 404 higher education students, 260 undergraduate and 144 graduate (specialization), who completed courses in the area of computer security, both deepening Course (Minor) Information Security and Computer Forensics of University Foundation of San Martin headquarters Puerto Colombia (FUSM) in the city of Barranquilla, as in the Specialization in Information Security offered at the University Research and Development (UDI) in the city of Bucaramanga.

Tests were conducted in the period since the second semester of 2006 to the second of 2010. The staff profile evaluated, mostly, graduates engineers and candidates for degree in computer engineering. A minority, less than 2%, constitutes electronics engineers and other professions. Table 4 shows the classification by level of education and time period:

TABLE 4. **DESCRIPTION OF PARTICIPANTS**

FUSM (degree)		UDI (graduate)	
PERIOD	AMOUNT	PERIOD	AMOUNT
2006-2	53	2008-1	20
2007-1	45	2008-2	25
2007-2	36	2009-1	24
2008-1	21	2009-2	27
2008-2	22	2010-1	23
2009-1	19	2010-2	25
2009-2	27	Total	144
2010-1	20		
2010-2	17		
Total	260		

Data obtained through the instrument

With the use of OEASegInf is found that besides providing a virtual environment for the evaluation motivates the development of competence placing difficult problems are solved according to the experience, this is described in Table 5.

TABLE 5. **SKILLS DEVELOPMENT**

Period	Institution	No. students	Average exams	Average progress in right questions	Average difference	Average percentage
2006-2	FUSM	53	7,5	4,49-7,43	2,94	65%
2007-1	FUSM	45	6,0	2,28-5,95	3,67	161%
2007-2	FUSM	36	6,5	2,00-6,88	4,88	244%
2008-1	FUSM	21	6,4	5,19-6,61	1,42	27%
2008-1	UDI	20	7,0	4,80-7,05	2,25	47%
2008-2	UDI	25	6,0	2,48-6,00	3,52	142%
2008-2	FUSM	22	5,0	3,77-6,41	2,64	70%
2009-1	FUSM	19	5,2	3,00-6,78	3,78	126%
2009-1	UDI	24	4,5	3,25-6,90	3,65	112%
2009-2	FUSM	27	2,1	2,92-5,18	2,26	77%
2009-2	UDI	27	3,5	3,18-6,67	3,49	110%
2010-1	FUSM	20	3,4	3,15-5,43	2,28	72%
2010-1	UDI	23	6,4	2,87-6,00	3,13	109%
2010-2	FUSM	17	5,3	2,88-5,53	2,65	92%
2010-2	UDI	25	6,6	3,04-6,32	3,28	108%
Average		**27**	**5,4**	**3,29-6,34**	**3,06**	**93%**

ANALYSIS OF THE SKILLS

Analyzing the data presented in Table 5, indicates that students feel safe to have done a good test when repeated, on average, 5 times (5.4), achieving real progress almost doubled in correct answers to pass an average of 3.29 to 6.34 representing an increase of 93%. In conjunction, the performance happens something similar as it goes from 36.5% (3.29 / 9) to 70.5% (6.34 / 9) during the reporting period.

This shows a significant advance in the development of interpretive and argumentative skills, especially the skills to interpret and articulate concepts. More specifically in the techniques and countermeasures applied to incidents. This is shown in Figure 3.

Also, is observed a tendency to an improvement in the progress of correct responses on a level about double. This is seen at a general level in the students, whether undergraduate or specialization. This trend is shown more clearly in Figure 4.

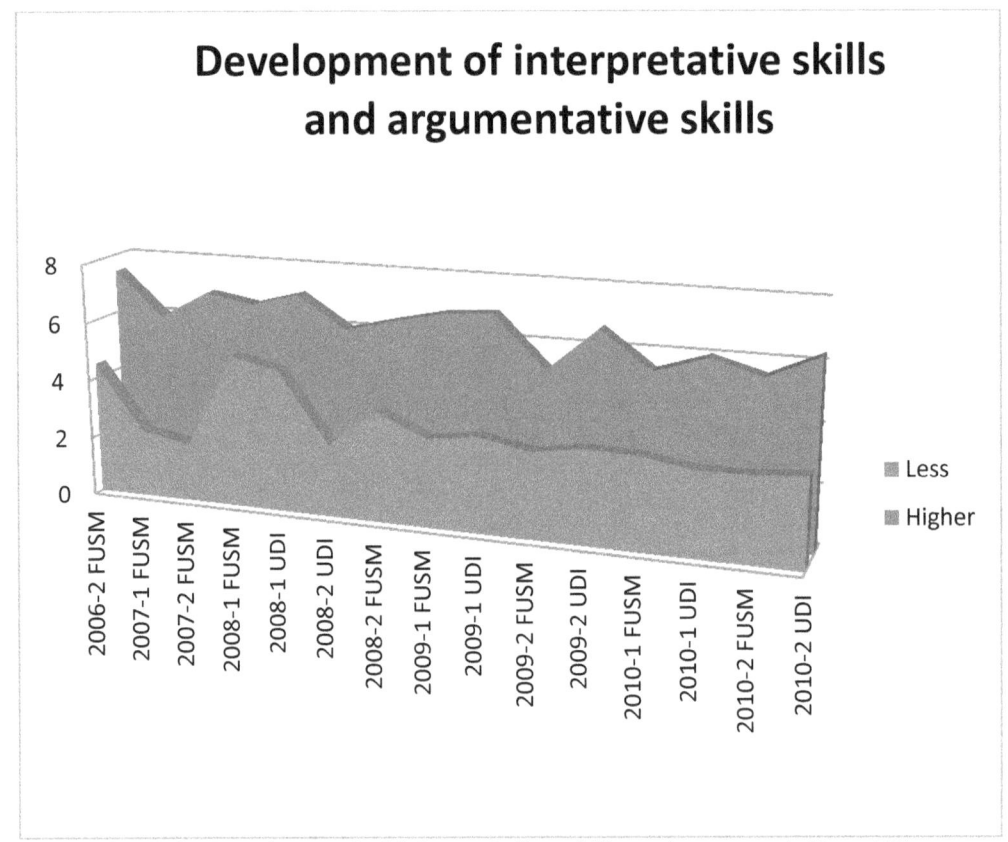

FIGURE 3. Development of interpretative skills and argumentative skills

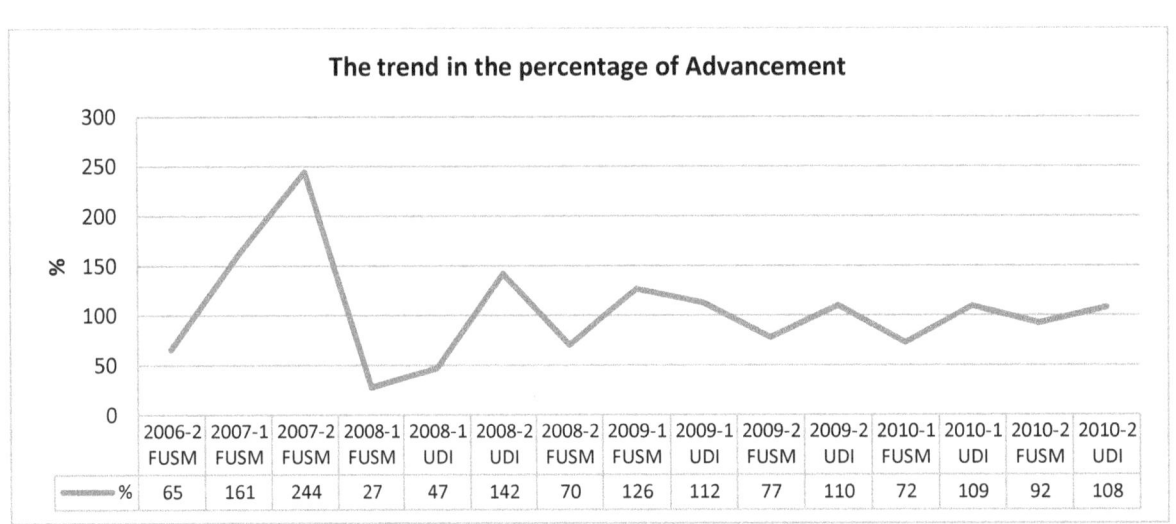

	2006-2 FUSM	2007-1 FUSM	2007-2 FUSM	2008-1 FUSM	2008-1 UDI	2008-2 UDI	2008-2 FUSM	2009-1 FUSM	2009-1 UDI	2009-2 FUSM	2009-2 UDI	2010-1 FUSM	2010-1 UDI	2010-2 FUSM	2010-2 UDI
%	65	161	244	27	47	142	70	126	112	77	110	72	109	92	108

FIGURE 4. Percentage of progress of all students

Skills at the undergraduate level

Taking only the data of students in the undergraduate level (Table 6), it appears that this type of student tends to do, too, the test 5 times (5.3) on average. Achieving group trend of progress in obtaining correct answers nearly double. Figure 5 shows the trend.

TABLE 6. **SKILLS AT THE UNDERGRADUATE LEVEL**

Period	Institution	No. students	Average exams	Average progress in right questions	Average difference	Average percentage
2006-2	FUSM	53	7,5	4,49-7,43	2,94	65%
2007-1	FUSM	45	6,0	2,28-5,95	3,67	161%
2007-2	FUSM	36	6,5	2,00-6,88	4,88	244%
2008-1	FUSM	21	6,4	5,19-6,61	1,42	27%
2008-2	FUSM	22	5,0	3,77-6,41	2,64	70%
2009-1	FUSM	19	5,2	3,00-6,78	3,78	126%
2009-2	FUSM	27	2,1	2,92-5,18	2,26	77%
2010-1	FUSM	20	3,4	3,15-5,43	2,28	72%
2010-2	FUSM	17	5,3	2,88-5,53	2,65	92%
Average		**29**	**5,3**	**3,30-6,24**	**2,95**	**89%**

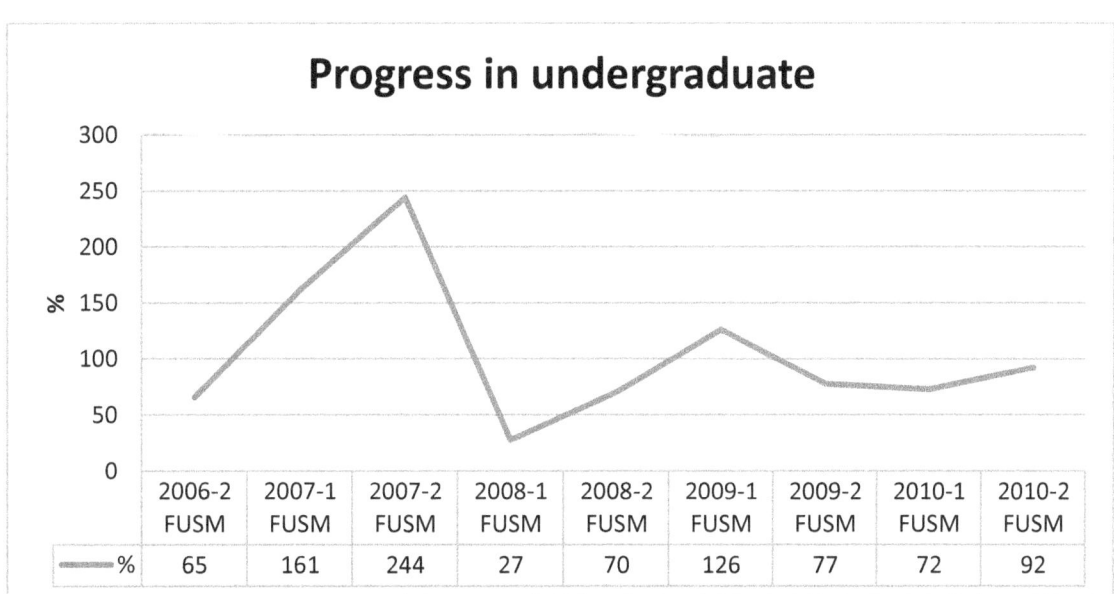

FIGURE 5. **Percentage of progress in undergraduate**

Skills at the graduate level

The behavior of students of the specialization is very similar to the undergraduate and the general. It highlights trends in developments close to 93%. The data for students of specialization are shown in Table 7 progress and trends are illustrated in Figure 6.

TABLE 7. **SKILLS AT THE GRADUATE LEVEL**

Period	Institution	No. students	Average exams	Average progress in right questions	Average difference	Average percentage
2008-1	UDI	20	7,0	4,80-7,05	2,25	47%
2008-2	UDI	25	6,0	2,48-6,00	3,52	142%
2009-1	UDI	24	4,5	3,25-6,90	3,65	112%
2009-2	UDI	27	3,5	3,18-6,67	3,49	110%
2010-1	UDI	23	6,4	2,87-6,00	3,13	109%
2010-2	UDI	25	6,6	3,04-6,32	3,28	108%
Averages		**27**	**5,4**	**3,29-6,34**	**3,06**	**93%**

FIGURE 6. **Percentage of progress in graduate**

Analysis of the Propositive skills

This project seeks to **break paradigms** from the propositive skills development in students. It aims to train skilled engineers in computer security capable provide solutions to problems differently, using practical learning.

In strut analysis, highlights the high attempt, about 63% of students, of **screenshots** of the problems in order to eliminate the time variable and solve the problem with accessible literature sources. In Table 8 details the summary of the screenshots.

TABLE 8. **ANALYSIS OF SCREENSHOTS**

Period	Institution	No. Students	Screenshots
2007-1	FUSM	45	3/45 = 7%
2007-2	FUSM	36	30/36 = 84%
2008-1	FUSM	21	21/21 = 100%
2008-1	UDI	20	17/20 = 85%
2008-2	UDI	25	20/25 = 80%
2008-2	FUSM	22	22/22 = 100%
2009-1	FUSM	19	12/19 = 63%
2009-1	UDI	24	15/24 = 63%
2009-2	FUSM	27	23/27 = 85%
2009-2	UDI	27	27/24 = 100%
2010-1	FUSM	20	10/20 = 50%
2010-1	UDI	23	17/23 = 74%
2010-2	FUSM	17	15/17 = 88%
2010-2	UDI	25	21/25 = 84%
		Totals	**253/404 = 62.6%**

By type of student, there is a difference of 29 percentage points in the preference of this tool, students of specialization on the undergraduate students. It is detailed in Table 9.

TABLE 9. **SCREENSHOTS BY LEVEL**

Type of student	No. Students	Screenshots	Percentage
Undergraduate	260	136	52.3%
Graduate - Specialization	144	117	81.3%

In general terms the screen capture is the most popular among students, both undergraduate and specialization as a way to extend the time to answer each question as more than half of students (63%) have used it. See Figure 7.

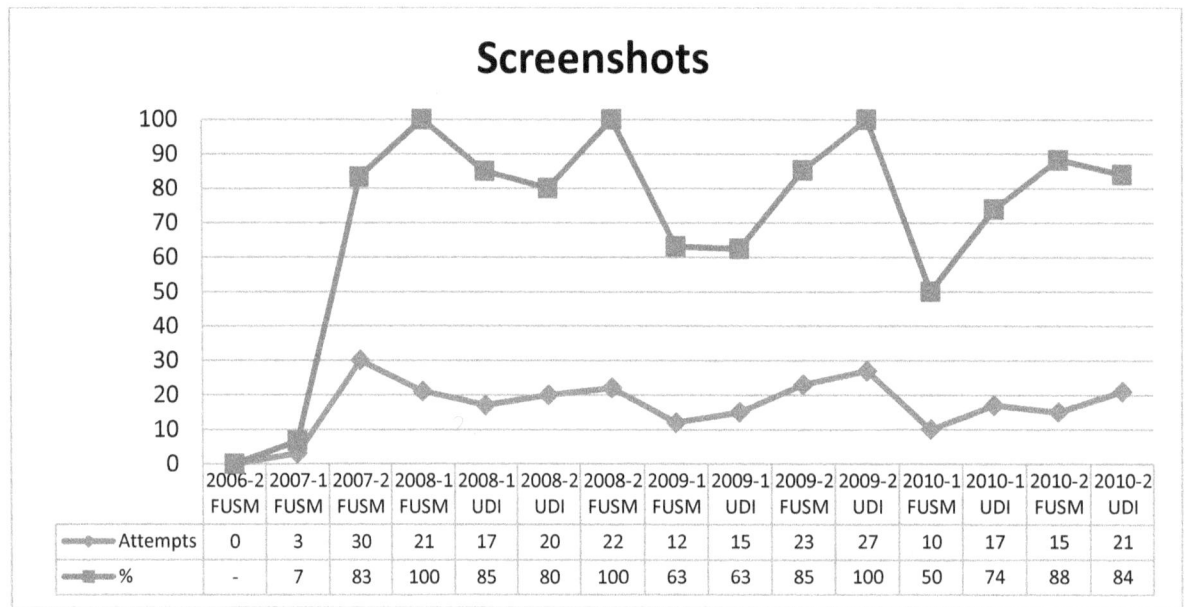

FIGURE 7. Screenshots

The technique of the **Software Cracking** to Learning's Evaluative Object has been little initiative only 20 unsuccessful attempts have been detected. These failed attempts represent less than 5% of the students, and presumably his low popularity is the product of self-imposed restrictions of students or simply did not occur at the time.

TABLE 10. **SOFTWARE CRACKING**

Period	Institution	No. Students	Unsuccessful software cracking
2008-1	UDI	20	2/20 = 10%
2010-1	UDI	23	7/23 = 30%
2010-2	FUSM	17	5/17= 29%
2010-2	UDI	25	6/25 = 24%
		Totals	**20/404 = 4.95%**

Additionally, originated from the comments received in feedback from students about the cracking of Learning's Evaluative Object saying that this option does not foresee either by the same stress generated by answering the exam or the technical complexity of the tool

Similarly, it find that the students need much experience and mathematical foundation for implementing a process of **cryptanalysis** the file generated by Learning's Evaluative Object. Perhaps for this reason, this technique has low rates of application, and has only been introduced during 2007. This is shown in Table 11.

TABLE 11. **CRYPTANALYSIS**

Period	Institution	No. Students	Cryptanalysis (unsuccessful)
2007-1	FUSM	45	1/45 = 2%
2007-2	FUSM	36	2/36 = 5%
2010-1	UDI	23	2/23 = 9%
2010-2	FUSM	17	2/17 = 12%
2010-2	UDI	25	3/25 = 12%
		Totals	**10/404 = 2.4%**

It is worth show in the present study, although the Learning's Evaluative Object not drive, **social engineering** attempts that students have done, having as the victim to the teacher, on the one hand, and the other, with the victim the students have already been tested. With respect to social engineering applied to the teacher, it is observed that the amount realized is greater than the cryptanalysis and software cracking, but rather low and infrequent. However, one attempt at social engineering to be successful enough so that all students obtain the necessary information and the experiment is not given naturally. Fortunately none of the attempts at social engineering has been successful thanks to the techniques used anti-social engineering by the teacher. In Table 12 shows an overview of the events.

TABLE 12. **SOCIAL ENGINEERING**

Period	Institution	No. Students	Social engineering
2007-1	FUSM	45	5/45 = 11%
2007-2	FUSM	36	3/36 = 8%
2008-1	FUSM	21	4/21 = 19%
2008-2	FUSM	22	2/22 = 9%
2009-1	FUSM	19	4/19 = 21%
2009-2	FUSM	27	2/27 = 7%
2009-2	UDI	27	3/27 = 11%
2010-1	UDI	23	10/23 = 43%
2010-2	FUSM	17	8/17 = 47%
2010-2	UDI	25	1/25 = 44%
		Totals	**52/404 = 12.9%**

It also notes that social engineering is more desired by undergraduate student by the graduate student (specialization)with one difference of about 5 percentage points. Table 13 details this observation.

TABLE 13. **SOCIAL ENGINEERING BY LEVEL**

Type of student	No. Students	Social engineering	Percentage
Undergraduate	260	38	14.6%
Graduate - Specialization	144	14	9.7%

In Figure 8 shows graphically the behavior of the attempts of social engineering attack having as the victim to the teacher.

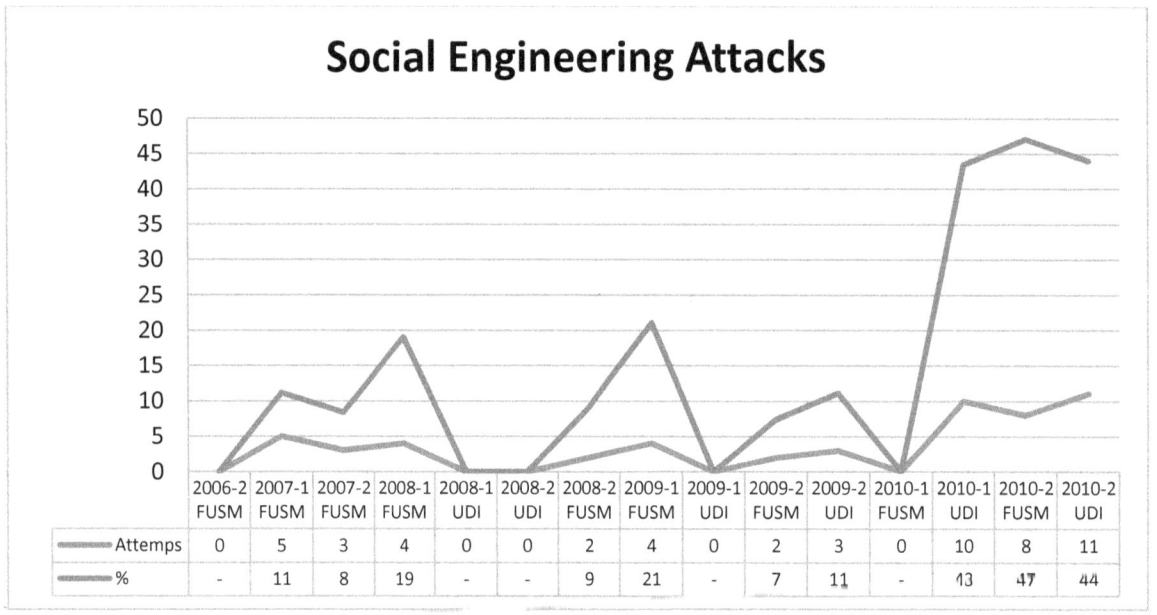

Social Engineering Attacks

	2006-2 FUSM	2007-1 FUSM	2007-2 FUSM	2008-1 FUSM	2008-1 UDI	2008-2 UDI	2008-2 FUSM	2009-1 FUSM	2009-1 UDI	2009-2 FUSM	2009-2 UDI	2010-1 FUSM	2010-1 UDI	2010-2 FUSM	2010-2 UDI
Attemps	0	5	3	4	0	0	2	4	0	2	3	0	10	8	11
%	-	11	8	19	-	-	9	21	-	7	11	-	13	47	44

FIGURE 8. Social Engineering Attacks

A curious situation has occurred front of social engineering should be applied to students who have already been tested, it happens that new students, who must do it, not think to ask their peers, which coursed the course, on this issue and regret not having done so, when it is too late.

Finally, among the results obtained showed a special case where an engineer who was studying specialization in Computer Security in the 2008-II was able to "stop time" installing the Learning's Evaluative Object on an operating system of a virtual machine and pausing execution of the virtual operating system whenever necessary, and thus could have enough time to find the answer through different media (Review articles, search the Internet, asking experts, etc..) to adequately resolve each problem.

Calculation of Confidence Intervals

Statistical to be used

To calculate confidence intervals for the number of exam carried out students, we use the formula for confidence interval estimation of the mean with unknown mean population [Berenson, 1996, Section 10.3. p. 334] based on the Student's t distribution. The raw material data for calculation are listed in Table 14.

TABLE 14. **DATA FOR THE CALCULATION OF CONFIDENCE INTERVALS**

Type of student	Average exams	Standard Deviation	Size
Undergraduate	5.267	1.655	9
Graduate - Specialization	5.667	1.367	6
All	5.426	1.508	15

The statistic to be used is as follows:

Sample Mean $\pm$ t$_{n-1}$ (standard deviation / (sample size) ^0.5)

Calculating the confidence interval of the General Population

It has the following:

n = 15; X_{Todos} = 5.426; T_{14} = 2.1448 (with a confidence level of 95% and 14 degrees of freedom); S_{Todos} = 1.508

Applying the formula: $X_{Todos} \pm T_{14} * S_{Todos}$ / (n)^ 0.5

= 5.426 $\pm$ (2.1448) * 1.508/ (15) ^ 0.5

= 5.426 $\pm$ 3.324 / 3.783

= 5.426 $\pm$ 0.835

Then the confidence interval is as follows: **4.591 <= U$_{Todos}$ <= 6.261.** This confidence interval is described in the following Figure 9:

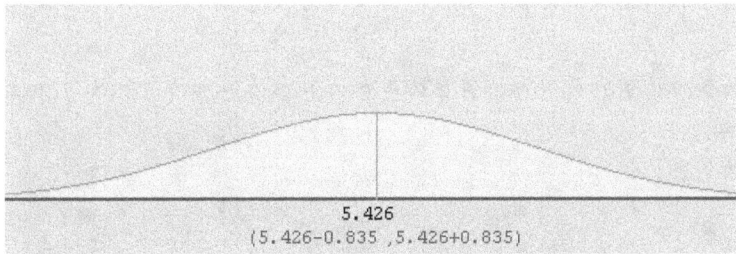

FIGURE 9. **Confidence interval of the General Population**

Calculating the confidence interval for undergraduate students

It has the following:

n = 9; $X_{Undergraduate}$ = 5.267; T_8 = 2.3060 (with a confidence level of 95% and 8 degrees of freedom); $S_{Undergraduate}$ = 1.655

Applying the formula: $X_{Undergraduate} \pm T_8 * S_{Undergraduate} / (n)^{0.5}$

= 5.267 ± (2.3060) * 1.655/ (9) ^ 0.5

= 5.267 ± 3.816 / 3.000

= 5.267 ± 1.272

Then the confidence interval is as follows: **3.994 <= U**$_{Undergraduate}$ **<= 6.539** This confidence interval is described in the following Figure 10:

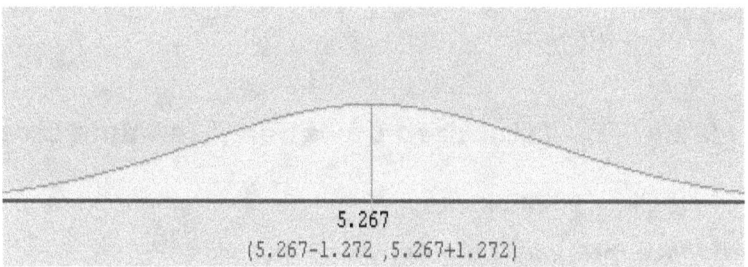

5.267
(5.267-1.272 ,5.267+1.272)

FIGURE 10. Confidence interval for undergraduate students

Calculating the confidence interval for graduate students

It has the following:

n = 6; $X_{Graduate}$ = 5.667; T_5 = 2.5706 (with a confidence level of 95% and 5 degrees of freedom); $S_{Graduate}$ = 1.367

Applying the formula: $X_{Graduate} \pm T_5 * S_{Graduate} / (n)^{0.5}$

= 5.667 ± (2.5706) * 1.367/ (6) ^ 0.5

= 5.667 ± 3.514 / 2.449

= 5.667 ± 1.4345

Then the confidence interval is as follows: **4.2324 <= U**$_{Graduate}$ **<= 7.1015** This confidence interval is described in the following Figure 11:

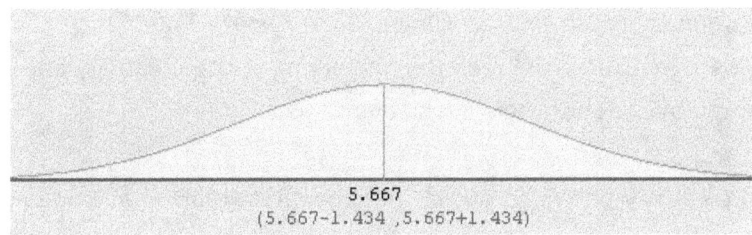

FIGURE 11. Confidence interval for undergraduate students

Hypothesis Testing

This section presents the arguments needed to verify the possibility of rejecting the hypothesis of this project. Remember the statement of the hypothesis raised earlier, namely:

The motivation for the use of creativity through the use of Learning's Evaluative Objects influences the development of skills and abilities that enable the student to propose nontraditional solutions to problems in the Computer Security area.

Next, consider the following arguments:

1. The 62.6% of students who participated in the experiment made use the **screen capture** technique as a means to master time, within the scenario of Learning's Evaluative Object. In this technique the graduate students exceeded by 29 percentage points to for the undergraduate students in reference to their use.

2. The technique of **software cracking**, performed at Learning's Evaluative Object, has been not show much initiative as only 4.95% of the students were encouraged to use it, because, by his own comments, self-imposed restrictions and complexity of the technique.

3. It was observed in the experiment that the process of **cryptanalysis** applied to the file generated for Learning's Evaluative Object is had an incidence of 2.4% due to its complexity.

4. Notable was in the experiment of **social engineering** attempts by students to the teacher (victim), It participation in the experiment is represented 12.9% of the students.

5. The **use of a virtual machine** that was able to "stop time" of the execution of Learning's Evaluative Object in the second semester of 2008 that left him helpless until then.

Of these five (5) arguments can be concluded that about 83.85% of the students involved (approximately 339) were motivated to break the paradigms at use creativity and it learned in class to solve the problems posed by the Learning's Evaluative Object.

All this outpouring of creativity described above, has the effect that on average students achieve real progress almost doubled in at correct answers from an average of 3.29 to 6.34, representing an increase of 93%, ie performance goes from 36.5% to 70.5%.

Finally, with these arguments, ends the test of hypothesis, enunciating that **CANNOT BE REJECTED** the hypothesis formulated for this project, and therefore it is claimed that:

The motivation for the use of creativity through the use of Learning's Evaluative Objects influences the development of skills and abilities that enable the student to propose nontraditional solutions to problems in the Computer Security area.

CONCLUSIONS AND FUTURE WORK

The Learning's Evaluative Objects represent an alternative to improve learning processes and especially in the assessment. This as a result of continuous monitoring of activities of the student to respond the problems, which can guide the teacher in the first instance, to know what resources used by students, and second, to improve the appropriation of knowledge and skills from results.

The results obtained using the Learning's Evaluative Object (OEASegInf) on the topic of computer security are interspersed. On the one hand, interpretative and argumentative skills (corresponding to the ability to know and know-how) it see significant progress as the student solves problems nearly doubled; and on the other hand, corresponding to the propositive skills (doing), it find techniques very popular and not so popular at the time of use. In fact, nearly 84% of students have used some technique to increase the time to answer questions or to circumvent the security of the Learning's Evaluative Object, which indicates a tendency to use what they learned in class in about two thirds of students on average. Therefore we can affirm that the Learning's Evaluative Object stimulates a significant degree the use of procedures and techniques that develop the competencies that are to be evaluated.

The structure proposed in this work of Learning's Evaluative Objects, Engine generator problems or questions, and Engine skills assessment, offers the possibility of reuse in different courses on the subject of security, without any kind of update in short periods of time, contrary to what happens when we generate static questionnaires. Therefore, this work provides the opportunity to expand into other areas of knowledge

Among the future work include: design and develop of the various Learning's Evaluative Objects with focus to assess other issues of engineering systems and observe the behavior of the results. Then it is necessary to extend the above purpose to other areas of knowledge.

Improve the evaluation object used to adjust to new attack techniques, this product of a risk analysis (threat + vulnerability) and to monitor new forms of risks.

BIBLIOGRAPHY

[Alexander, 2007] Alberto G. Alexander. 2007. Diseño de un sistema de gestión de seguridad de información. Óptica ISO 27001:2005. Alfaomega Grupo Editor. 2007. 176 páginas: ISBN: 9586827133

[Arsham, 1995] Arsham, H. 1995. Interactive education: Impact of the internet on learning & teaching. DOI=http://UBMAIL.ubalt.edu/harsham/interactive.htm. Visitada el 12/03/2010

[Baryamureeba and Tushabe, 2004] Venansius Baryamureeba and Florence Tushabe. 2004. The Enhanced Digital Investigation Process Model. Disponible en la red: http://citeseerx.ist.psu.edu/viewdoc/download?doi=10.1.1.60.492&rep=rep1&type=pdf. Visitado: 26/11/2011

[Berenson, 1996] Berenson, Mark and Levine, David. (1996) Estadística básica en administración: Conceptos y aplicaciones.4 Ed. Prentice – Hall, México. 946 p.

[Caballero Pino, 2003] Caballero Pino, G. 2003. Introducción a la Criptografía. 2 Edición. Alfaomega Ra-Ma. México.

[Cano, 2009] Jeimy J. Cano M. Computación forense. Descubriendo los rastros informáticos. Alfaomega Grupo Editor, S.A. de C.V. (México, D.F.) 2009. ISBN: 9789586827676. páginas: 329

[Casey, 2009] Eoghan Casey. Handbook of Digital Forensics and Investigation. ISBN-10: 0123742676 | ISBN-13: 978-0123742674 | Publication Date: November 9, 2009

[Díaz, Montero, & Aedo, 2005] Díaz, M, Montero, S & Aedo, I. 2005. Ingeniería Web y patrones de diseño. Universidad Carlos III Madrid. Prentice – Hall, Madrid. 409 p.

[Durán, 2010] Fernando Durán. Seguridad Informática en la Empresa: Teoría y Práctica de Seguridad para Empleados y Gerentes No Técnicos. 2010

[Friesen, 2001] Friesen, N. 2001. What are educational objects? Interactive learning environments, Vol. 9, No. 3, pp. 219-230.

[Hadnagy, 2011] Christopher Hadnagy. Ingenieria social / Social engineering: El Arte Del Hacking Personal / the Art of Hacking. Anaya Multimedia (June 30, 2011). ISBN-10: 8441529655. 400 pages

[Hurtado Carmona, 2011b] Dougglas Hurtado Carmona, "Teoría General de sistemas: un enfoque hacia la ingeniería de sistemas" En: Colombia 2011. ed:Lulu.com Enterprises ISBN: 978-1-257-78193-5 pags. 125

[Hurtado Carmona, 2011c]

Dougglas Hurtado Carmona, "General System Theory: A focus on computer science engineering" En: Colombia 2011. ed:Lulu.com Enterprises ISBN: 978-1-257-78224-6 pags. 126

[Hurtado Carmona, 2011a]

Dougglas Hurtado Carmona, "Análisis del desarrollo de competencias desde la enseñanza asistida por computador" En: Colombia 2011. ed:Lulu.com Enterprises ISBN: 978-1-257-81753-5 pags. 44

[Hurtado Carmona, 2011d]

Dougglas Hurtado Carmona, "Analysis of skills development from computer-assisted teaching" En: Colombia 2011. ed:Lulu.com Enterprises ISBN: 978-1-257-81756-6 pags. 46

[Hurtado Carmona, 2010]

Dougglas Hurtado Carmona, "Desarrollo de competencias en seguridad informática a partir de objetos evaluativos del aprendizaje" En: Colombia. 2010. Evento: X Jornada Nacional de Seguridad Informática ACIS 2010 Ponencia:Desarrollo de competencias en seguridad informática a partir de objetos evaluativos del aprendizaje Disponible en: http://www.acis.org.co/fileadmin/Base_de_Conocimiento/ X_JornadaSeguridad/ArticuloDouglasHurtado.pdf

[JOHANSEN, 1996]

JOHANSEN B, Oscar. Introducción a la teoría general de sistemas, – Decimotercera reimpresión - Noriega Editores, 1996.

[Johnsonbaugh, 2005]

Johnsonbaugh, R. 2005. Matemáticas dicretas.Sexta edición. Pearson Education. México. 696 pag.

[Kevin Mitnick, 2006]

Diario el país – España 25/06/2006. Reportaje – Tecnología: Los mejores consejos de un 'superhacker', entrevista otorgada por Kevin Mitnick.

[Long, 2004]

Johnny Long. Hacking con Google/ Hacking with Google (Hackers Y Seguridad / Hackers and Security) (Spanish Edition). Anaya Multimedia; 3ra edition (June 30, 2005)ISBN-10: 8441518513. 508 pages

[Maiorano y Fernández, 2009]

Ariel Maiorano, Damián Fernández. Criptografía. Técnicas de desarrollo para profesionales. Alfaomega Grupo Editor, S.A. de C.V. (México, D. F.).2009. ISBN: 9789872311384. páginas: 276

[Mann, 2011]

Mik Mann. 2011. Seguridad Informatica (Spanish Edition). Kindle Edition - Oct 13, 2011 - Kindle eBook

[McClure et al, 2010]

McClure, Stuart, Scambray, Joel, Kurtz. George. 2010. Hackers 6: secretos y soluciones de seguridad de redes. McGraw Hill. México. 688 Pág.

[NIJ, 2001]

National Institute of Justice. (July 2001) Electronic Crime Scene Investigation A Guide for First Responders https://www.ncjrs.gov/pdffiles1/nij/187736.pdf. Fecha de consulta: Noviembre 29 de 2011.

[NIJ, 2008]

National Institute of Justice.(April 2008) Electronic Crime Scene Investigation A Guide for First Responders http://www.nij.gov/publications/ecrime-guide-219941/

[Sanz, Aedo, y Díaz, 2006]

Sanz, Daniel, Aedo, Ignacio y Díaz, Paloma 2006. Un Servicio Web de Políticas de Acceso Basadas en Roles para Hipermedia.

DOI=http://www.ewh.ieee.org/reg/9/etrans/vol4issue2Ap
ril2006/4TLA2_3Sanz.pdf. Visitada el 24/06/2009

[Vitturini et al, 2005]

Vitturini, M., Benedetti, L., y Señas, P. 2005. Filtros de corrección automática como objetos de aprendizaje evaluativos para sistemas educativos basados en la web. DOI=http://cs.uns.edu.ar/lidine/publicaciones/FCA%20co mo%20objetos%20de%20aprendizaje%20evaluativos%2 0para%20SEBW.pdf. Visitada el 14/06/2007

[Wiley, 2000]

Wiley, David. 2000. Learning Object Design and Sequencing Theory. Tesis doctoral no publicada de la Brigham Young University. DOI=http://davidwiley.com/papers/dissertation/dissertati on.pdf. Visitada el 24/06/2009

[Wiley, 2001]

Wiley, D. 2001. Connecting learning objects to instructional design theory: A definition, a methaphor, and a taxonomy.

[Wiley, 2006]

Wiley, D. 2006 R.I.P. ping on Learning Objects DOI= http://opencontent.org/blog/archives/230   Visitada   el 14/06/2007

[Yong-Dal Shin. 2008]

Yong-Dal Shin. 2008. New Digital Forensics Investigation Procedure Model. This paper appears in: Networked Computing and Advanced Information Management, 2008. NCM '08. Fourth International Conference on
Page(s): 528 - 531. Issue Date: 2-4 Sept. 2008. Volume: 1. 978-0-7695-3322-3/08 $25.00 © 2008 IEEE. DOI 10.1109/NCM.2008.116

[Zemanek, 2004]

Jakub Zemanek. Cracking sin Secretos: Ataque y defensa de software. RA-MA EDITORIAL. 2004. 391 Páginas

www.ingramcontent.com/pod-product-compliance
Lightning Source LLC
Chambersburg PA
CBHW081357170526
45166CB00010B/3120